the Elements of th...

NICKEL

Aubrey Stimola

rosen
central

The Rosen Publishing Group, Inc., New York

For my dad, Michael, who always took me to the Museum of Natural History, taught me to fish with the barbs down, and still saves me the Tuesday Science Times

Published in 2007 by The Rosen Publishing Group, Inc.
29 East 21st Street, New York, NY 10010

Library of Congress Cataloging-in-Publication Data

Stimola, Aubrey.
Nickel / Aubrey Stimola.—1st ed.
      p. cm.—(Understanding the elements of the periodic table)
Includes bibliographical references and index.
ISBN 1-4042-0704-X (library binding)
1. Nickel—Popular works. 2. Periodic law—Popular works.
I. Title. II. Series.
QD181.N6S75 2007
546'.625—dc22

                                                    2005034813

Manufactured in the United States of America

On the cover: Nickel's square on the periodic table of the elements; the atomic structure of a nickel atom *(inset)*.

Contents

# Introduction

Long before it was officially identified as a unique substance, nickel (Ni) played an important role in many ancient cultures around the world. Bronzes containing nickel were used in what is now Syria as early as 3500 BC. Between 1700 and 1400 BC, the ancient Chinese developed and used a shiny, strong material called *pai-t'ung* (white copper) in their weapons, coins, and artwork. This valuable material was exported to the Middle East and as far as Europe. Scientists now believe that pai-t'ung was actually a mixture of nickel and other metals.

In the Middle Ages, blacksmiths used mixtures of nickel and other metals to make suits of armor, swords, and tools. Ancient Peruvians valued a shiny, corrosion-resistant metal that they believed was a kind of silver (Ag), but which many scientists believe to have been nickel. In Germany, a nickel-containing mineral was used to color glass green. Little did any of these ancients know that nickel was responsible for most of the properties they prized.

Chapter One
What Is Nickel?

Nickel is a hard, silvery white metal that is extremely shiny and malleable, or easily shaped and molded. It is also ductile, meaning it can be stretched into sheets or wires. It can withstand very high and very low temperatures, does not rust, and is one of only three naturally magnetic elements. As we will see, these qualities make nickel an extremely important element.

Nickel is the twenty-fourth most abundant element in Earth's crust. It can be found in all soils, in the ocean floors, and in ocean water itself. Nickel is also emitted during volcanic eruptions. The biggest source of nickel on the planet, however, is unreachable because it is buried deep in Earth's molten core. Scientists believe that nickel makes up approximately 7 to 10 percent of the planet's core.

The Devil's Copper

In the 1700s, copper miners in the Saxony region of Germany discovered a strange substance that was slightly lighter in color than copper (Cu). Unlike copper, this mysterious ore turned a bright silver color when refined, rather than reddish brown. Also unlike copper, it was extremely hard. Because no use could be found for this uncooperative material other than to give glass a greenish hue, copper miners believed that the devil had

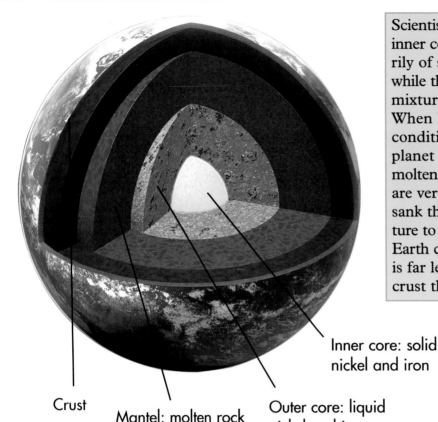

Scientists theorize that Earth's inner core is composed primarily of solid nickel and iron while the outer core is a molten mixture of these elements. When Earth formed, extreme conditions caused the new planet to become entirely molten. Because nickel and iron are very heavy, these elements sank through the molten mixture to what became the core as Earth cooled. This is why nickel is far less abundant in Earth's crust than it is at the core.

Inner core: solid nickel and iron

Crust

Mantel: molten rock

Outer core: liquid nickel and iron

planted the substance to deceive them in their search for real copper. As a result, it became known as *Kupfernickel*, or "the devil's copper."

The substance that gave the ore its unique characteristics was finally identified in 1751. Mineralogist Baron Axel Fredrik Cronstedt was investigating a mineral called niccolite that had been found in a mine in Sweden. Like the German miners years before, Cronstedt was expecting to extract copper from this interesting ore, but his experiments repeatedly resulted in the extraction of a hard, shiny, whitish metal. Realizing that further attempts to extract copper from Kupfernickel would be fruitless, the frustrated Cronstedt named this new metal "nickel," the Swedish word for "Old Nick," or the devil.

At first, the chemists of the time were unwilling to accept that a new element had been discovered. They were sure that nickel was really a mixture of many other already known substances, such as iron (Fe), cobalt (Co), arsenic (As), and copper. It wasn't until 1775 that another Swedish chemist, Torbern Bergman, extracted from niccolite a sample of nickel so pure that it proved beyond a shadow of a doubt that the new metal was not a combination of other substances but its own unique element.

It's Elemental

An element is something that is made of only one kind of atom. When Cronstedt attempted to extract copper from niccolite, his experiments repeatedly yielded the strange, shiny, white metal from which he could extract nothing else. This inability to be broken down into simpler components is what makes nickel an element. Each element is made up of its own special atoms. For example, the element nickel is made up entirely of nickel atoms, just as the element iron is made up entirely of iron atoms.

Because they are the most basic of all substances, elements are the building blocks of everything on Earth. Elements can join together in many different ways to create all the things you see around you. The air you breathe, the water you drink, even the book you are reading are all composed of a combination of various elements.

Just as puzzle pieces can be put together to make a picture, elements combine to make the range of substances in the universe by fitting together in specific ways. Different substances are made depending upon which elements fit together. For example, when an atom of the element oxygen (O) combines with two atoms of the element hydrogen (H), a water molecule is formed. Substances like water that are made up of more than one type of atom are called compounds.

Subatomic Particles

Atoms themselves are made up of protons, neutrons, and electrons. These are known as subatomic particles.

Discovering Subatomic Particles

The first person to discover the existence of subatomic particles was physicist J. J. Thomson, who identified the electron in 1897. Thomson knew that atoms must also have a positive charge to balance out the negative charge of electrons. This positive charge was supplied by protons, which were identified in 1919 by the physicist Ernest Rutherford. Neutrons were discovered even later, in 1932, by the physicist James Chadwick.

Protons and neutrons are found in the atom's center, or nucleus. Because protons have a positive electrical charge and neutrons have no charge at all, the nucleus of an atom is always positively charged. Electrons circulate around the nucleus, filling most of the space in the atom. The electrons are organized in shells. An atom can have several electron shells, and each shell can hold a certain number of electrons.

Because electrons have negative electrical charges, they are drawn to the positive charges of the protons in the atom's

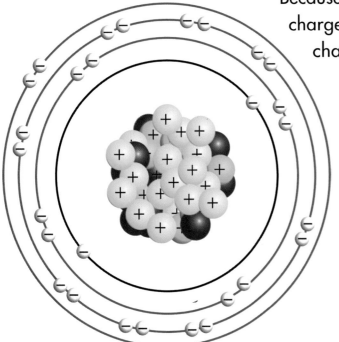

A nickel atom has twenty-eight protons and thirty neutrons in its nucleus. Twenty-eight negatively charged electrons orbit the nucleus in layers called electron shells and are attracted to the positive charges of the protons. Nickel atoms are fairly stable, meaning that they are neither particularly reactive, nor particularly unreactive.

nucleus. This attraction is similar to the way a magnet is attracted to a refrigerator door. The attraction between electrons and protons, known as the electromagnetic force, is what holds atoms together.

Since electrons have negative charges, they tend to stay as far away from other electrons as possible. The number of negatively charged electrons in an atom always matches the number of positively charged protons (the atomic number). As a result, electrons and protons cancel each other out, leaving the atom with no overall charge.

An Atom of Nickel

The number of electrons an atom has and how they are arranged in shells are important in determining how the atom will behave. Generally speaking, when an atom has a filled outer shell of electrons, it is stable and does not react. When the outermost shell is not full, the atom is unstable and therefore reactive. In order to fill its outermost shell, an unstable atom will react with other atoms to either borrow or share electrons.

Metals, Semimetals, and Nonmetals

Over three quarters of the elements known to man are metals, including nickel. The others fall into one of two categories. Some, such as germanium (Ge), boron (B), and silicon (Si), are known as semimetals, or metalloids. There are seven generally accepted naturally occurring semimetals. The remaining elements are nonmetals, such as carbon (C), sulfur (S), and iodine (I). There are only sixteen naturally occurring nonmetals.

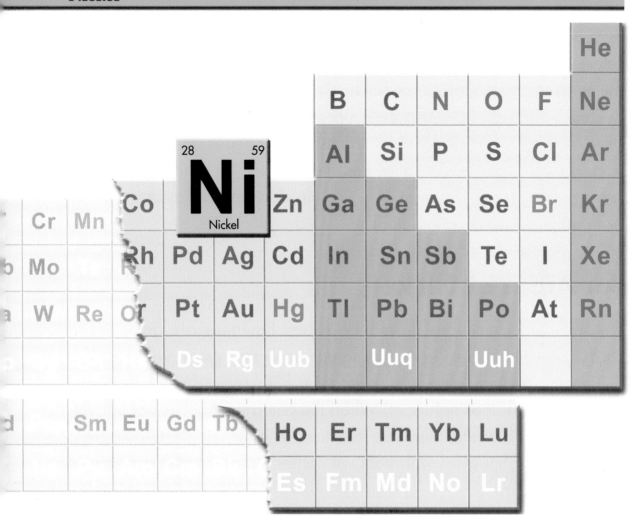

The number in the upper left-hand corner of nickel's square on the periodic table is its atomic number, or the number of protons in its nucleus. The number in the upper right-hand corner is its atomic weight, or the average mass of the element's isotopes. Isotopes are atoms of an element that have the same number of protons in their nucleus as other atoms of the same element, but a different number of neutrons. The mass of each isotope is the sum of all protons and neutrons in its nucleus. If a nickel isotope has twenty-eight protons and thirty-six neutrons, its atomic number is twenty-eight and it has an atomic weight of sixty-four.

Because we know that a nickel atom has twenty-eight protons, we also know it has twenty-eight electrons. Generally, when an atom's first electron shell is filled, any additional electrons spill over into the next shell. When the second shell is filled, remaining electrons go into the third shell, and so on. Moving outward from the nucleus, the first three electron shells of a typical atom are full when they contain 2, 8, and 18 electrons, respectively. But when the third shell is less than half full, two electrons enter the fourth shell, and only then does the third shell continue to fill. You can understand how this happens by remembering the rows of seats at the last movie theater you went to. Did you notice how sometimes people prefer to sit in the back rows even if the rows in front are not full? The two electrons in a nickel atom's outer shell act in the same manner. This electron arrangement is responsible for nickel's unique properties.

Nickel atoms seldom share these electrons with other atoms. As we will see, nickel's resistance to sharing or giving up the electrons in its outer shell is part of what makes nickel a very useful and versatile element. At the same time, however, the fact that nickel's third electron shell has only sixteen electrons, instead of eighteen, enables nickel atoms to take part in some interesting and complex chemical reactions.

Nickel and Its Compounds

Because of their particular electron arrangement, nickel atoms can react with atoms of other elements to form a number of different compounds. Compounds are substances made up of chemically combined atoms of two or more elements. Almost all chemical reactions involving nickel result in each nickel atom losing the two electrons in its outer shell. These electrons combine with atoms of different elements that need them to become stable.

When an atom loses electrons, it is no longer electrically neutral. When a nickel atom loses its outer two electrons, it becomes a nickel ion with a charge of +2. This is because the nickel ion now has two more positively charged protons than it does negatively charged electrons. Likewise, if a single atom should get the two electrons lost by nickel, it would become an ion with a −2 charge because it now has two more negatively charged electrons than positively charged protons. Because these two ions have opposite charges, they are attracted to each other and an ionic bond forms between them. Together, they are considered an ionic compound. Although a nickel atom can exist as an ion with charges ranging from −1 to +4, its most stable ion form is +2.

Because nickel atoms have an inner electron shell that is not quite full, they can form a great number of different compounds. These compounds are generally very complex and enable nickel to play very important roles in the world.

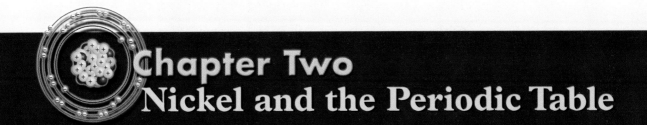

Chapter Two
Nickel and the Periodic Table

There are 116 different elements known to humans, 90 of which occur naturally. Scientists have created the remaining 26 elements in laboratories. It is possible that there are still some undiscovered elements.

It wasn't until 1869 that a Russian chemist by the name of Dmitry Mendeleyev designed a method of categorizing the elements. Mendeleyev began by looking for patterns among the various elements and found that some of them shared similar chemical and physical properties such as density and melting points. Next, he wrote down all the known characteristics of each element on note cards and tried to organize them in different ways. When the elements were arranged in horizontal rows in order of increasing atomic weight, he noticed that patterns began to emerge.

Mendeleyev's organization of the elements by atomic weight did have some flaws, the largest of which was that it resulted in a lot of gaps between the known elements. The Russian chemist boldly proposed that the blanks actually represented elements that had not yet been discovered. He also predicted some of the characteristics of these yet undiscovered elements based on the location of the gaps in what he called the periodic table of the elements. (The word "periodic" refers to something that has

Typische Elemente							
			K = 39	Rb = 85	Cs = 133	—	—
			Ca = 40	Sr = 87	Ba = 137	—	—
			—	?Yt = 88?	?Di = 138?	Er = 178?	—
			Ti = 48?	Zr = 90	Ce = 140?	?La = 180?	Tb = 231
			V = 51	Nb = 94	—	Ta = 182	—
			Cr = 52	Mo = 96	—	W = 184	U = 240
			Mn = 55	—	—	—	—
			Fe = 56	Ru = 104	—	Os = 195?	—
			Co = 59	Rh = 104	—	Ir = 197	—
			Ni = 59	Pd = 106	—	Pt = 198?	—
H = 1 Li = 7	Na = 23	Cu = 63	Ag = 108	—	Au = 199?	—	
Be = 9,4	Mg = 24	Zn = 65	Cd = 112	—	Hg = 200	—	
B = 11	Al = 27,3	—	In = 113	—	Tl = 204	—	
C = 12	Si = 28	—	Sn = 118	—	Pb = 207	—	
N = 14	P = 31	As = 75	Sb = 122	—	Bi = 208	—	
O = 16	S = 32	Se = 78	Te = 125?	—	—	—	
F = 19	Cl = 35,5	Br = 80	J = 127	—	—	—	

The current periodic table of the elements is organized quite differently from the original version shown here. This is due to the discovery of new elements, as well as a greater understanding of previously discovered elements and how they relate to and react with each other. As you can see, Mendeleyev left many gaps in his periodic table for elements he correctly theorized would be discovered later. An element discovered in 1955, atomic number 101, was named mendelevium (Md) in his honor.

repeating patterns.) Less than twenty years after the 1869 publication of the periodic table, scientists had filled three of the gaps with newly discovered elements. These new elements fit into the periodic table exactly where Mendeleyev predicted they would.

Since its first publication, the periodic table of the elements has undergone many changes, including the discovery and addition of more than fifty elements. However, Mendeleyev's method of organization continues to be the basis for our modern periodic table.

The elements are arranged in horizontal rows called periods by atomic number, or number of protons. They are also arranged into eighteen

Nickel Snapshot

Chemical Symbol:	Ni
Classification:	Transition metal
Properties:	Magnetic, malleable, ductile
Discovered By:	Identified by mineralogist Baron Axel Fredrik Cronstedt in 1751
Atomic Number:	28
Atomic Weight:	58.693 atomic mass units (amu)
Protons:	28
Electrons:	28
Neutrons:	usually 30
Density at 68°F (20°C):	8.908 grams per cubic centimeter (g/cm³)
Melting Point:	2,647°F (1,453°C)
Boiling Point:	5,275°F (2,913°C)
Commonly Found:	Earth's crust, Earth's core, ocean floor

Almost an Element

It's not easy for a substance to be added to the list of accepted elements and to earn a spot on the periodic table. In 1999, scientists at the U.S. Department of Energy thought they had created atoms of new elements 116 (ununhexium) and 118 (ununoctium) by smashing krypton ions into lead using a machine called a cyclotron. While element 116 has been re-created, no other experiments were able to duplicate the creation of element 118. As a result, in 2001, scientists took back the claim of having discovered this new element.

First built in the early 1930s, a cyclotron is a type of particle accelerator. Particle accelerators are used to create very high-speed, high-energy beams of subatomic particles, which then collide with target atoms. The smaller particles and energy that result from these intense collisions are then analyzed in detectors. One reason scientists are interested in particle acceleration is because it provides opportunities to discover new forms of matter as well as what smaller parts make up matter.

vertical columns, or groups, numbered IA through VIIA, IB through VIIIB, and 0. Just as members of a family have similarities, the elements within each group have similar chemical properties. In fact, elements in the same group are often referred to as being part of the same family.

Tossing Nickel on the Table

So where does nickel fit in the periodic table? Nickel is part of group VIIIB (group 10 in some versions), which is part of a larger group of elements known as transition metals. Transition metals can be described as being typical metals. Many of them have important uses in manufacturing and industry. As we will learn, nickel is particularly important in this regard.

Properties of Nickel

Properties are the physical and chemical characteristics that make an element unique. Like many of its neighboring transition metals, nickel is strong, hard, shiny, easy to shape, a fairly good conductor of heat and electricity, and less reactive than some other metals. Nickel also has a high melting point and a high boiling point. All elements on the periodic table can exist in one or more of three physical states, or phases: solid, liquid, and gas. Nickel is a solid at room temperature but will become liquid when heated to 2,647 degrees Fahrenheit (1,453 degrees Celsius). It will boil and become a gas at 5,275°F (2,913°C).

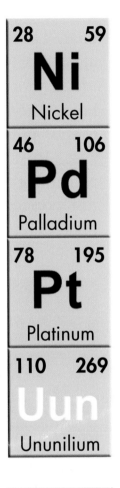

Nickel belongs to a large family of elements known as transition metals. Transition metals are noted for the unique arrangement of their outermost electrons, which causes them to behave differently than other metals and gives them unique physical and chemical properties. Nickel is found in the first period of transition metals and is part of a smaller group known as VIIIB. Other transition metals you may be familiar with are iron, copper, zinc (Zn), and titanium (Ti).

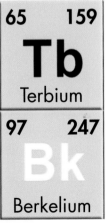

The Effects of Corrosion

Corrosion is a chemical reaction that takes place between metal and oxygen in the presence of moisture. This reaction is called oxidation. When nickel corrodes, its surface becomes coated with a layer of oxide. This process is what causes metals like silver to tarnish, or lose their shine. In the case of nickel, this oxide layer actually protects the metal's atoms below the surface from further corrosion. Iron is a very strong metal, but it corrodes easily, resulting in the formation of rust. Rust causes metals to become brittle. To create rust resistance, iron is often coated or mixed with metals that do not corrode easily, such as nickel. In fact, the oxidation of nickel creates a coating on the surface of metals that actually protects them from further corrosion.

Some metals do not form protective oxide coatings. When iron corrodes, a flaky layer of rust forms. Rust can lift away, allowing the metal beneath to corrode.

However, nickel is one of only three elements, along with iron and cobalt, that are ferromagnetic. Ferromagnetic substances are those that can be permanently magnetized. When magnetized, they are attracted to iron and a few other metals, and make excellent materials for magnets. Nickel is also very ductile, meaning it can be stretched and hammered into thin wires and sheets. It is also quite malleable, or easily molded. Another important characteristic of nickel is its resistance to corrosion, or breaking down when exposed to oxygen in the air.

Nickel Isotopes

While most nickel atoms have twenty-eight protons and thirty neutrons, some nickel atoms have extra neutrons. These forms of the element are called nickel isotopes. Many elements, including nickel, have more than one isotope.

You can think of isotopes as different forms of the same atoms. Neutrons have mass but no electrical charge, so nickel isotopes have only slightly different physical properties, such as atomic weight. There are five stable isotopes of nickel. The official atomic weight of nickel (approximately fifty-nine atomic mass units) is the average of the mass of all five nickel isotopes.

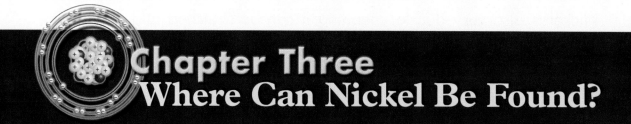

Chapter Three
Where Can Nickel Be Found?

We have already learned that nickel is the twenty-fourth most abundant element in Earth's crust and that 7 to 10 percent of Earth's molten core is made up of nickel. But how did it get there?

Earth is thought to have formed billions of years ago when a large number of particles drew together into a cloud. The cloud grew increasingly dense as it attracted more and more particles. Eventually, this massive cloud condensed into a liquid. Heavier substances like nickel and iron sank and collected in the center of the newly formed planet, creating its core.

Much of the nickel that is not buried deep within Earth's core is thought to have arrived in the form of asteroids, meteorites, and space dust. Throughout Earth's 4.5-billion-year lifetime, the planet has been pummeled by space debris and meteorites. In fact, scientists believe that the world's biggest nickel deposit, which is located in Canada, formed when an enormous asteroid containing nickel struck Earth almost 2 billion years ago. Besides containing a lot of nickel, scientists believe that the asteroid's forceful impact also resulted in volcanic eruptions. During these eruptions, a great deal of nickel was expelled from Earth's core.

Identifying Meteorites

There are three different types of meteorites: stony, iron, and stony-iron. The percentage of nickel a meteorite contains helps distinguish which category it belongs in. The most common meteorites are of the stony variety and contain 15 to 25 percent nickel-iron. Iron meteorites, the second most common type of meteorite, may be composed almost entirely of a nickel-iron combination.

Seen here is a stony-iron meteorite, also known as a pallasite. Some scientists theorize that these meteorites had their origins in a rocky planet that broke apart long ago. Most scientists, however, believe these meteorites to be left-over debris from when Earth and other planets were formed.

Where Nickel Is Found

Nickel is rarely found in nature in its pure form. Instead, it is locked away in mineral ores, such as the niccolite that frustrated German miners looking for copper. Nickel is generally found in sulfide ores called pyrrhotite and pentlandite. It is also found in complex ores known as laterites. All of these ores must first be extracted from the earth and then purified in refineries in order to make use of the nickel within.

Slag, a waste product created in the production of nickel, is dumped at a nickel foundry in western Russia. Nickel is made into usable items at foundries. This process, called metal casting, involves pouring the molten nickel (or a mixture of nickel and other metals) into a mold for a desired product, such as a propeller for a boat. After the metal cools, solidifies, and is removed from the mold, the mold can be used over and over again.

Large-scale nickel mining began in 1876. A number of different methods are used to extract nickel from the earth. When nickel ores are buried deep in the ground, tunnels are dug in the earth and the nickel ore is excavated and transported back up to the surface. When nickel ores are found closer to the surface, explosives are used to blast them out. Once removed from mines, the nickel ores are sent to refineries, where pure nickel is separated from the rest of the minerals. This is precisely the task that Cronstedt undertook when he tried to break down "the devil's copper" into its most basic, elemental components. Modern technology has since made this process far easier than it was for poor Cronstedt.

The nickel refining process begins by pulverizing the ores to create a fine powder. For the sulfide ores, pyrrhotite and pentlandite, this powder is mixed with water. Air bubbles are introduced into the water. The particles containing nickel stick to the air bubbles, which are then carried to the surface where they form a thick film. This film is scraped off and dried, resulting in a powder of pure nickel sulfide. Repeating this process several times ensures that as many of the nickel-containing particles as possible have been captured.

The nickel sulfide powder is then smelted—a high-temperature process that causes pure molten nickel to sink and waste products, or slag, to float to the top where they are removed. This process, too, is repeated several times, resulting in a nickel sample that is almost 100 percent pure.

For complex laterite ores, different processes are used to refine pure nickel. These multistep processes often involve binding nickel to sulfides in order to use refining methods similar to those used for refining pure nickel from pyrrhotite and pentlandite ores.

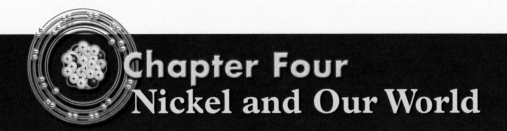

Chapter Four
Nickel and Our World

Nickel plays very important roles in our everyday lives, even though most people are unaware of it. Because of its unique properties, nickel is used in building materials, batteries, water purification, chemical production, food preparation, machinery, household appliances, and transportation.

The Role of Pure Nickel

Because nickel is naturally resistant to the effects of oxygen and other chemicals, thin layers of pure nickel are often applied to the surfaces of metals that are not as resistant to corrosion. One process for accomplishing this is called electroplating.

Electroplating involves the attraction of positive nickel ions in a chemical solution to a negatively charged metal surface. The ions gain electrons from the negatively charged metal and form nickel atoms. These atoms create a uniform covering that shields the metal beneath from the corrosive effects of oxygen. Currently, scientists are even using electroplating to protect plastic surfaces. The plastic is first treated with a coat of solid metal, which is then electroplated with nickel. Electroplating is used to protect jewelry, computer disks, and kitchen and bathroom hardware.

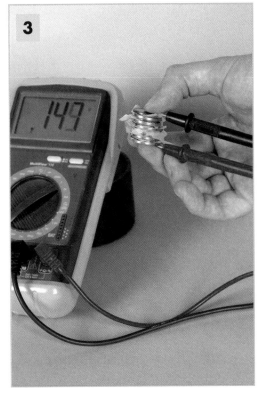

To make a coin battery, you'll need five pennies, five nickels, soap, a salt-water solution, paper napkins, and a voltmeter.

Directions:
1. After cleaning the coins with soap, soak the napkins in saltwater solution.
2. Tear the napkins into pieces slightly larger than the size of the coins, and make a "sandwich" by stacking a penny, a piece of napkin, a nickel, a piece of napkin, and so on.
3. Connect a voltmeter to the ends of the stack.

What's happening?
The saltwater solution is an electrolyte. The electrolyte reacts with metals in the coins, which act as electrodes. The two coins contain different metals (copper in the penny and a nickel-copper alloy in the nickel), one of which reacts more strongly than the other. This leaves an electrical potential difference between them. In batteries, this difference is called voltage.

The Role of Nickel Compounds

When the German miners unearthed copper-colored niccolite in their quest for true copper, they found that the uncooperative substance could be used to color glass green. The substance responsible for this green coloration was a compound called nickel arsenide. Similar nickel compounds are still used today in glass coloring and pottery glazes.

Rechargeable batteries require the use of an important nickel compound. Batteries harness energy released from contained chemical reactions and turn it into electrical energy that can be used to power devices like cell phones, radios, and even cars. All batteries have a positive end (cathode) and a negative end (anode) made from two different metals or metal compounds. An electrolyte solution between the cathode and anode allows chemical reactions to take place between them. This chemical reaction causes a buildup of electrons at the anode. Just as the negatively charged electrons in an atom are attracted to the positively charged protons in the atom's nucleus, the electrons at the anode are attracted to the positively charged cathode. However, the chemical reaction occurring in the battery keeps this from happening. When a battery is inserted into a device that requires power and the device is turned on, a closed circuit, or a path for the electrical energy to travel, is created. The electrons from the anode then travel through the device in order to reach the cathode. The electrons release energy into the device in the process.

In nickel-containing batteries, the cathode contains a compound called nickel oxyhydroxide. The anode can contain either a metal hydride or the element cadmium. Nickel-containing batteries are much more efficient and convenient than other kinds of batteries because they can be recharged. In other battery types, the chemicals involved in the reactions lose their ability to supply electrons, causing the battery to die. But the chemical reactions that take place in nickel-containing batteries

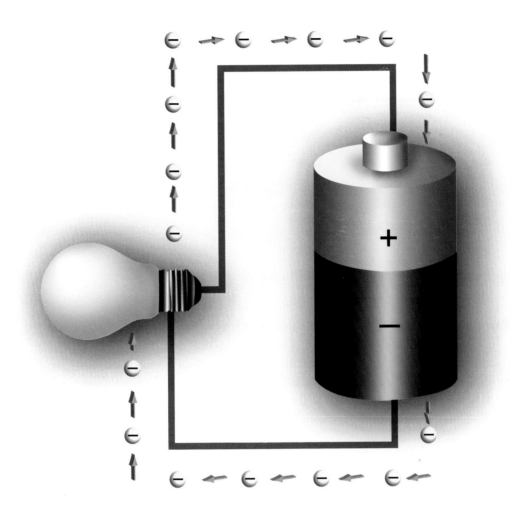

This diagram shows a battery powering a lightbulb. Electrons produced by chemical reactions inside the battery travel from the negatively charged anode along the wire, through the lightbulb, powering it. After passing through the lightbulb, the electrons enter the battery at the positively charged cathode. A battery "dies" when the chemicals involved in these reactions are used up.

can be reversed, restoring the battery to its original state, chock-full of stored energy.

Nickel has the ability to bond with large amounts of hydrogen gas to form nickel hydride. Hydrogen, when it reacts with oxygen, is a powerful source of pollution-free energy, but the reaction can be extremely dangerous

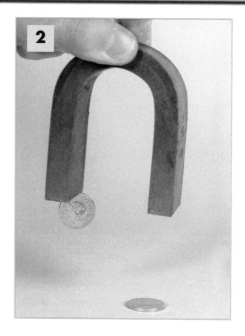

To test nickel's magnetic properties, take an American five-cent piece from 2005 and a Canadian five-cent piece minted between 1955 and 1981. Hold a magnet near each coin. The Canadian five-cent piece is more attracted to the magnet because it contains a high concentration of nickel. The American five-cent piece is only 25 percent nickel and is less attracted to the magnet.

if not carefully controlled and contained. Nickel and other metal hydrides could be the easiest and safest way to store hydrogen for use in energy-producing reactions. Scientists are working on ways of releasing the hydrogen bound in nickel hydrides in a way that would allow them to efficiently capture the most energy.

Nickel in Steel Production

Without question, the most important role nickel plays in our world is in the production of steel. Steel is a very strong metal that is widely used in industry and technology. Steel is used in the construction of everything from skyscrapers to cars to medical equipment. Steel itself is not an elemental metal, however. Instead, it is a mixture of metals, or an alloy.

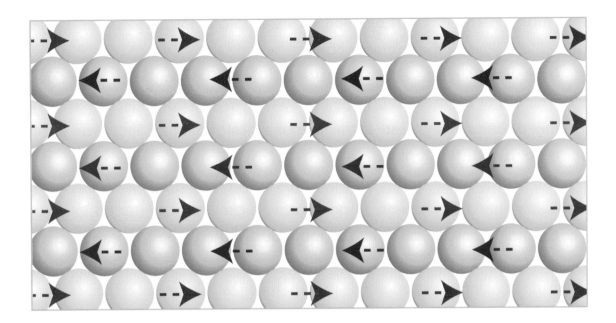

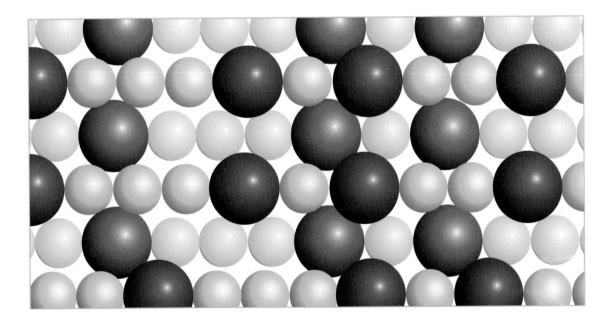

The small spheres in the top image represent atoms in a pure, elemental metal. Because the atoms in pure metals are identical and of only one type, the atoms fit together in neat rows that can slide over each other. For this reason, pure metals can become brittle and crack under certain conditions. However, when atoms of another metal are added—indicated by the larger spheres in the bottom image—the two types of atom can't line up as neatly or slide past each other. The resulting metal mixture is an alloy that is stronger than the original metals.

Alloys are valuable because they have different properties from the elements that make them up. Each element on its own cannot perform the functions that alloys of that element can.

Nickel-containing steels are resistant to corrosion and are very ductile. The use of these steels can be traced all the way back to 1889, when engineer James Riley used them to make armor plating that was both strong and heat resistant.

The properties of different kinds of nickel-containing steels depend on the combination of elements in addition to nickel that are added to iron. For example, some nickel steels are more ductile than others. These stretchy steels are used in structures that must give under tremendous pressure without breaking or snapping, such as bridges. Other nickel steels can withstand very low temperatures and are used in containers that store liquids with very low boiling points, such as nitrogen and hydrogen. There are also nickel steels prized for their extreme hardness; these are used in rocket fuel tanks.

Stainless Steel

Nickel is most commonly used in the production of stainless steel, a shiny, silvery alloy that is particularly resistant to corrosion. Stainless steel is mostly made of iron. It typically contains only 8 percent nickel, 20 percent chromium, and a small amount of other elements. The nickel and chromium combine to protect the surface of the steel.

Invar

Invar is a very unique nickel alloy made of 36 percent nickel and 64 percent iron. Unlike most metals, invar is resistant to expanding when heated and contracting when cooled. Because of its unique properties, invar is used in devices that undergo frequent and rapid temperature changes,

The Golden Gate Bridge in San Franciso, California, spans 1.7 miles (2.7 kilometers) and is made entirely of steel and concrete. Bridges must be able to expand and contract in order to withstand vertical and horizontal sway, as well as fluctuations in temperature and pressure due to weight and wind. They must also withstand corrosion. Because steels containing nickel are flexible, they are regularly used in structures like bridges and tall buildings.

such as teakettles, thermostats, and electrical components. Because it does not expand or contract, invar is also widely used in measurement devices that need to be precise, like metal tape measures.

Other Nickel Alloys

Nickel is used to make several other alloys besides steel. These alloys also combine the special properties of nickel with the special properties unique to other elements.

Monel

Monel is a nickel alloy made of approximately 70 percent nickel and 30 percent copper. Monel also contains small amounts of other metals. This nickel alloy is fairly hard and is very resistant to corrosion by salty seawater and other chemicals. These properties make it ideal for making propellers and other boat parts. Monel is also used in desalinization plants, which remove salt from water, and to coat cables that run underwater.

Expansion and Contraction

Temperature changes can make substances expand or contract. You may have noticed the effects of this kind of change around your house. Filling a cool glass with a hot beverage can cause the glass to break, as some parts of the glass contract more rapidly than others. Did you ever have trouble opening a glass jar? Running hot water over the lid will loosen it because the metal lid will expand more than the glass. Some nickel alloys are resistant to this expansion and contraction, which, over time, can cause a metal to fatigue and break.

Monel's resistance to other chemicals also makes it ideal for food processing and other industries.

Nichrome

Nichrome is made of nickel combined with 11 to 22 percent chromium (Cr) and small amounts of other elements. This incredible alloy remains strong even in red-hot heat, giving it a wide variety of uses.

Nichrome and other nickel-chromium alloys actually get stronger as their temperature increases. These alloys are often called superalloys.

All of the items above are transition metals. The ball bearings are made from nickel, the rod is made from iron, the necklace is made from silver, the reddish strip is made from copper, the black rock is manganese (Mn), and the gray lump is cobalt. Some transition metals are often used in their pure form, such as the copper that makes electrical wires. Others, like cobalt, nickel, and manganese, are mixed with other metals to make steels and other alloys.

Nickel in Coins

The United States was not the first country to use nickel in its coins. In fact, nickel was first used to make coins in 1860 in Belgium, and the first coin of pure nickel was made in 1881.

In the United States, five-cent coins (called half-dimes) were originally made of silver. After the American Civil War, however, there was a shortage of silver, which resulted in the decision to use a combination of copper and nickel. Copper alone bent too much and corroded easily. The same copper-nickel alloy is used in all U.S. coins today but in different amounts. For example, if you look at the edge of a U.S. quarter or dime, you'll notice layers of different colors. That's because only the faces of the coins are made out of a copper-nickel alloy; the insides are made of pure copper.

The appearance of the United States' five-cent coin, the nickel, has undergone many revisions over time. At one point nickels depicted an image of a Native American and an American bison. The nickels shown here are from a special series designed to commemorate the bicentennial of the Louisiana Purchase and the Lewis and Clark expedition of 1803.

Nichrome is commonly used in heating elements for toaster ovens and electric ovens; in spark plugs; and, perhaps most importantly, in jet engines, rocket engines, and turbines. Jet engines experience extreme heat and pressure. Turbines, devices powered by extremely hot steam to generate electricity, were at one time made of steel. Because steel cannot withstand the same high temperatures that nichrome can, these turbines broke more frequently than modern turbines.

Alloys of nickel and titanium display a very unique ability to "remember" their shape, even after being stretched and strained with great force and over long periods of time. These alloys are used today in many items, including eyeglass frames, coffeepots, space shuttles, and medical devices, and in robotics.

Nickel's magnetic properties make it useful in the production of magnets, particularly when alloyed with certain other elements. Alnico is the most important magnetic nickel alloy. It creates a very strong magnetic field that functions even at high temperatures. Alnico gets its name from combining the symbols of the elements from which it is produced: aluminum, nickel, and cobalt. It can be used in laboratory instruments, sensors, and communication devices.

Mu metals are another type of magnetic nickel alloy made from nickel, iron, and other elements. Mu metals are used to block magnetic fields created in laboratories and medical facilities.

A Helping Hand in the Lab

Aside from its roles in making rechargeable batteries and durable, useful metal alloys, nickel is also a common catalyst. A catalyst is something added to a chemical reaction that makes the reaction happen faster. Powdered nickel is often used to speed up a process called hydrogenation. Hydrogenation is the addition of atoms of hydrogen to liquid oils and fats to make them more solid. This process was once widely used to make soaps and is still used to make many different kinds of peanut butter. It is also used to make margarine, which is made out of hydrogenated vegetable oil. Though these reactions would occur without the presence of a catalyst, nickel makes the reaction happen faster.

Chapter Five
The Role of Nickel in Living Things

Scientists who study nutrition think that people need only very small amounts of nickel in their diets. They are still unsure, however, of exactly what role nickel plays in the human body.

Scientists have conducted experiments where they observe what happens to animals that are biologically similar to human beings when they don't get enough nickel. For example, when baby chicks and rats don't get enough nickel, they sometimes have liver problems. The liver is an important organ that plays many roles, including the production of substances that help in the digestion of food. Nickel may also be involved in growth, the production of breast milk, and helping our bodies make proteins. Foods such as walnuts, almonds, oatmeal, cocoa, soy beans, fresh and dried vegetables, peas, beans, and tea leaves contain large amounts of nickel.

Nickel has important functions in many plants. Nickel helps some plants absorb iron, it helps others to break down certain chemicals, and it may be necessary in the sprouting of seeds. Tea has more than twice the amount of nickel in its leaves than most other plants, and nickel also plays various roles in fungi and bacteria.

Sometimes people have allergic reactions to nickel from wearing inexpensive jewelry that is electroplated with nickel or nickel alloys. This allergy can develop over time, even after years of wearing the same kind of jewelry. This gradual process is called sensitizing. An allergic reaction

Many common food items contain some amount of nickel. These include oatmeal, beans, almonds, soybeans, walnuts, fresh vegetables, and tea. So far, scientists are unclear what role nickel plays in the human diet.

to nickel generally appears as a reddish, bumpy, and sometimes blistery rash that can be very itchy. It is often made worse by sweating. A reaction caused by something that irritates the skin is called contact dermatitis. The rash generally heals when the nickel-containing item is no longer worn. People with nickel allergies should wear only jewelry that is silver, gold (Au), or stainless steel, and they should avoid wearing items plated with nickel alloys against their skin.

In 2002, researchers in Europe discovered that some of the new coins produced by the European Union caused allergic reactions when held in the hands of a few very nickel-sensitive people. The coins in question contained two particular nickel alloys. Electrolytes in sweat from the hands of these nickel-sensitive people caused the coins to corrode faster than usual, releasing more nickel to the skin than other coins do.

Throughout this book, we have learned that nickel plays many important roles. Nickel gives us rechargeable batteries, which can be reused rather than thrown away. It makes flying around the world and even into space safer and easier. It makes bridges and buildings strong, and it even makes some important chemical reactions happen faster. Everywhere you go, you can bet that nickel is either right there on the surface or hiding just beneath it.

The Periodic Table of Elements

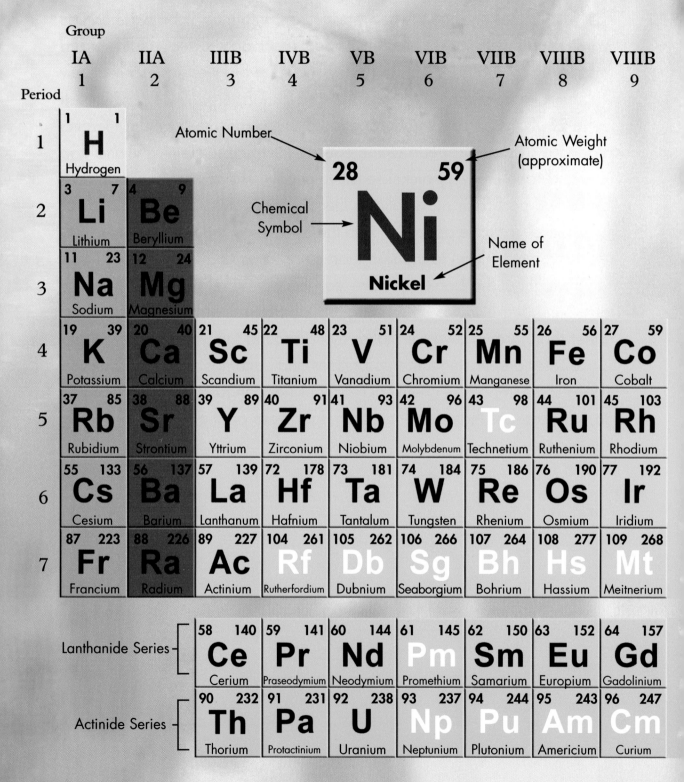

Group

IA	IIA	IIIB	IVB	VB	VIB	VIIB	VIIIB	VIIIB
1	2	3	4	5	6	7	8	9

Period

Atomic Number → 28

Atomic Weight (approximate) → 59

Chemical Symbol → Ni

Name of Element → Nickel

Period	IA	IIA	IIIB	IVB	VB	VIB	VIIB	VIIIB	VIIIB
1	1 / 1 **H** Hydrogen								
2	3 / 7 **Li** Lithium	4 / 9 **Be** Beryllium							
3	11 / 23 **Na** Sodium	12 / 24 **Mg** Magnesium							
4	19 / 39 **K** Potassium	20 / 40 **Ca** Calcium	21 / 45 **Sc** Scandium	22 / 48 **Ti** Titanium	23 / 51 **V** Vanadium	24 / 52 **Cr** Chromium	25 / 55 **Mn** Manganese	26 / 56 **Fe** Iron	27 / 59 **Co** Cobalt
5	37 / 85 **Rb** Rubidium	38 / 88 **Sr** Strontium	39 / 89 **Y** Yttrium	40 / 91 **Zr** Zirconium	41 / 93 **Nb** Niobium	42 / 96 **Mo** Molybdenum	43 / 98 **Tc** Technetium	44 / 101 **Ru** Ruthenium	45 / 103 **Rh** Rhodium
6	55 / 133 **Cs** Cesium	56 / 137 **Ba** Barium	57 / 139 **La** Lanthanum	72 / 178 **Hf** Hafnium	73 / 181 **Ta** Tantalum	74 / 184 **W** Tungsten	75 / 186 **Re** Rhenium	76 / 190 **Os** Osmium	77 / 192 **Ir** Iridium
7	87 / 223 **Fr** Francium	88 / 226 **Ra** Radium	89 / 227 **Ac** Actinium	104 / 261 **Rf** Rutherfordium	105 / 262 **Db** Dubnium	106 / 266 **Sg** Seaborgium	107 / 264 **Bh** Bohrium	108 / 277 **Hs** Hassium	109 / 268 **Mt** Meitnerium

Lanthanide Series

58 / 140 **Ce** Cerium	59 / 141 **Pr** Praseodymium	60 / 144 **Nd** Neodymium	61 / 145 **Pm** Promethium	62 / 150 **Sm** Samarium	63 / 152 **Eu** Europium	64 / 157 **Gd** Gadolinium

Actinide Series

90 / 232 **Th** Thorium	91 / 231 **Pa** Protactinium	92 / 238 **U** Uranium	93 / 237 **Np** Neptunium	94 / 244 **Pu** Plutonium	95 / 243 **Am** Americium	96 / 247 **Cm** Curium

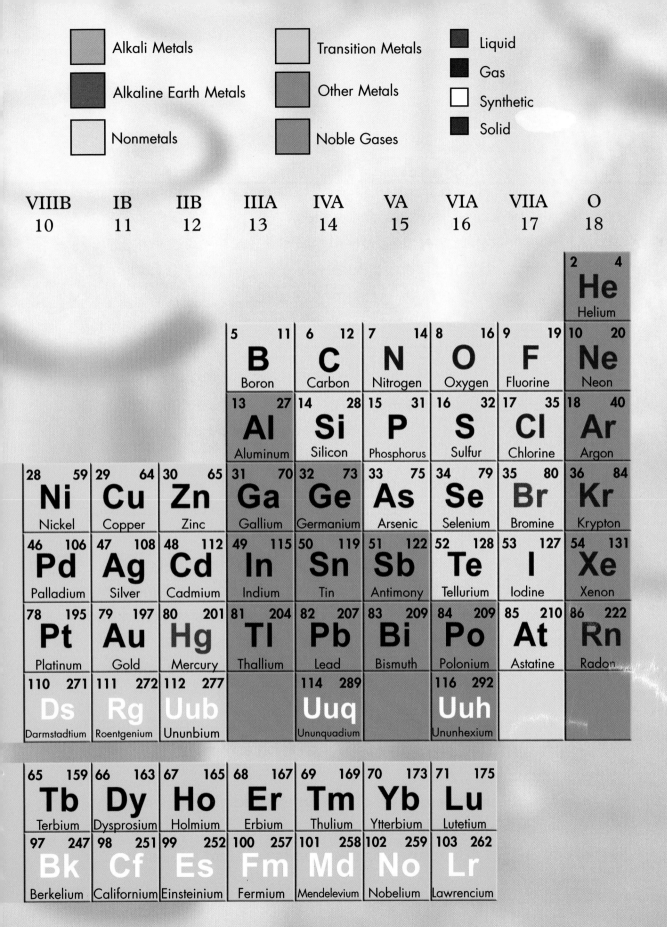

Legend:
- Alkali Metals
- Alkaline Earth Metals
- Nonmetals
- Transition Metals
- Other Metals
- Noble Gases
- Liquid
- Gas
- Synthetic
- Solid

| VIIIB | IB | IIB | IIIA | IVA | VA | VIA | VIIA | O |
| 10 | 11 | 12 | 13 | 14 | 15 | 16 | 17 | 18 |

								2　4 He Helium
			5　11 B Boron	6　12 C Carbon	7　14 N Nitrogen	8　16 O Oxygen	9　19 F Fluorine	10　20 Ne Neon
			13　27 Al Aluminum	14　28 Si Silicon	15　31 P Phosphorus	16　32 S Sulfur	17　35 Cl Chlorine	18　40 Ar Argon
28　59 Ni Nickel	29　64 Cu Copper	30　65 Zn Zinc	31　70 Ga Gallium	32　73 Ge Germanium	33　75 As Arsenic	34　79 Se Selenium	35　80 Br Bromine	36　84 Kr Krypton
46　106 Pd Palladium	47　108 Ag Silver	48　112 Cd Cadmium	49　115 In Indium	50　119 Sn Tin	51　122 Sb Antimony	52　128 Te Tellurium	53　127 I Iodine	54　131 Xe Xenon
78　195 Pt Platinum	79　197 Au Gold	80　201 Hg Mercury	81　204 Tl Thallium	82　207 Pb Lead	83　209 Bi Bismuth	84　209 Po Polonium	85　210 At Astatine	86　222 Rn Radon
110　271 Ds Darmstadtium	111　272 Rg Roentgenium	112　277 Uub Ununbium		114　289 Uuq Ununquadium		116　292 Uuh Ununhexium		

| 65　159 Tb Terbium | 66　163 Dy Dysprosium | 67　165 Ho Holmium | 68　167 Er Erbium | 69　169 Tm Thulium | 70　173 Yb Ytterbium | 71　175 Lu Lutetium |
| 97　247 Bk Berkelium | 98　251 Cf Californium | 99　252 Es Einsteinium | 100　257 Fm Fermium | 101　258 Md Mendelevium | 102　259 No Nobelium | 103　262 Lr Lawrencium |

Glossary

alloy A mixture of two or more metals, or a metal and a nonmetal.

anode The negative electrode of a battery.

atom The smallest part of an element that still has the properties of that element. Each atom has a positively charged nucleus and negatively charged electrons. Its net charge is zero.

atomic mass unit A unit of mass that is approximately equivalent to one atom of hydrogen.

atomic number The number of protons in the nucleus of an atom.

atomic weight The average mass of all isotopes of an element.

battery A source of electrical potential energy made up of two or more electrochemical cells.

cathode The positive electrode of a battery.

corrosion The process by which the surface of a metal is worn away by oxygen and chemicals.

cyclotron A type of particle accelerator in which charged particles are accelerated while confined to a circular path by a magnetic field.

ductile A term used to describe substances that can be stretched and hammered into thin wires and sheets.

electrode A conductor through which an electrical current enters or leaves.

electron A negatively charged particle that orbits the nucleus of an atom.

electroplating The process of covering an object with a thin layer of metal using electrolysis.

element A substance made up of one type of atom that cannot be broken down into simpler substances.

ferromagnetic A term used to describe metals that are easily magnetized.

hydrogenation The addition of hydrogen atoms to a molecule.

ion An atom that has become charged by losing or gaining electrons.

isotope Two or more forms of atoms of the same element that have the same atomic number but different atomic weights. This is because they have the same number of protons in their nuclei but varying numbers of neutrons. Isotopes of a single element possess almost identical chemical properties.

malleable A term used to describe metals that are easily shaped and molded.

matter Anything that has mass and occupies space.

meteorite A meteor that has fallen through the atmosphere and landed on Earth.

mineral Any of the naturally occurring compounds of which rocks and ores are made.

mining The process of removing minerals from the ground.

molecule The smallest particle of an element, consisting of one or more atoms, that exists on its own and still maintains its properties.

neutron A subatomic particle with no charge that is found in the nucleus of an atom.

niccolite A mineral that consists of 43.9 percent nickel and 56.1 percent arsenic.

nucleus The core of an atom that contains protons and (except in the case of hydrogen) neutrons.

ore A mineral from which useful substances like metals can be extracted.

proton A subatomic particle with a positive charge that is found in the nucleus of an atom.

refining The process of removing minor impurities from metal ores.

smelting The process of extracting a metal from an ore by repeatedly heating to high temperatures and removing impurities.

turbine A machine with blades attached to a central shaft. The pressure of water or steam on these blades causes the turbine to spin.

American Chemical Society
1155 16th Street NW
Washington, DC 20036
Web site: http://www.acs.org

Nickel Institute
55 University Avenue, Suite 1801
Toronto, ON M5J 2H7
Canada
Web site: http://www.nickelinstitute.org

Web Sites

Due to the changing nature of Internet links, the Rosen Publishing Group, Inc., has developed an online list of Web sites related to the subject of this book. This site is updated regularly. Please use this link to access the list:

http://www.rosenlinks.com/uept/nick

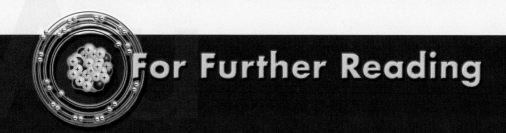

For Further Reading

Herr, Norman, and James Cunningham. *Hands-On Chemistry Activities with Real-Life Applications: Easy-to-Use Labs and Demonstrations for Grades 8–12.* New York, NY: Jossey-Bass, 2002.

Robinson, Tom. *The Everything Kids' Science Experiments Book: Boil Ice, Float Water, Measure Gravity; Challenge the World Around You!* Avon, MA: Adams Media Corporation, 2001.

Bibliography

Atkins, P. W. *The Periodic Kingdom: A Journey into the Land of the Chemical Elements.* New York, NY: Basic Books, 1997.

Emsley, John. *Nature's Building Blocks: An A–Z Guide to the Elements.* New York, NY: Oxford University Press, 2002.

Flyvholm, M. A., et al. "Nickel Content of Food and Estimation of Dietary Intake." *Z Lebensm Unters Forsch*, December 1984, Vol. 179, No. 6, pp. 427–431.

Heiserman, David L. *Exploring Chemical Elements and Their Compounds.* New York, NY: McGraw-Hill, 1991.

Morrison, R. T., and R. N. Boyd. *Organic Chemistry.* 6th ed. San Francisco, CA: Benjamin Cummings, 1992.

Rogers, Kirsteen, et al. *The Usborne Internet-Linked Science Encyclopedia.* London, England: Usborne Publishing, 2003.

Sparrow, Giles. *Nickel.* New York, NY: Benchmark Books, 2005.

Index

About the Author

Specializing in the translation of academic science into language comprehensible by and compelling to non-scientists, writer Aubrey Stimola has covered a variety of science topics. She has a degree in bioethics from Bard College and lives in Brooklyn, New York.

Photo Credits

Cover, pp. 1, 8, 10, 17, 27, 29, 40–41 by Tahara Anderson; p. 6 © Roger Harris/Photo Researchers, Inc.; p. 14 © SPL/Photo Researchers, Inc.; p. 16 © CC Studio/Photo Researchers, Inc.; p. 18 © Robert J. Erwin/Photo Researchers, Inc.; p. 21 © Erich Schrempp/Photo Researchers, Inc.; p. 22 © B & C Alexander/Photo Researchers, Inc.; pp. 25, 28 by Maura McConnell; p. 31 © Galen Rowell/Corbis; p. 33 © Andrew Lambert Photography/Photo Researchers, Inc.; p. 34 © Mark Wilson/Getty Images; p. 38 (cookie) © Lew Robertson/Corbis; p. 38 (beans, soybeans, vegetables) © Photo Disc/Market Fresh; p. 38 (peas, walnuts) © Photo Disc/Fruits and Vegetables; p. 38 (almonds) © Digital Stock Food and Ingredients; p. 38 (tea) © O. Lassen/zefa/Corbis.

Special thanks to Megan Roberts, director of science, Region 9 Schools, New York City, NY, and Jenny Ingber, high school chemistry teacher, Region 9 Schools, New York City, NY, for their assistance in executing the science experiments illustrated in this book.

Designer: Tahara Anderson